BIG FOUR CAMERAMAN

R.J. Blenkinsop

PRC

First published in the U.K. by Oxford Publishing Co.

This edition published by The Promotional Reprint Company Ltd
exclusively for Bookmart Limited, Desford Road, Enderby
Leicester LE9 5AD.

ISBN 1 85648 321 5

Printed in China

Preface

This new omnibus edition is a combination of four books which have been out of print for some years. The object of the photographs was to portray some of the Regional train workings of British Railways with roughly an equal number of pictures from the Eastern, Midland, Southern and Western Regions.

The geographical selection was not difficult as my travels in the 1950s did not cover the whole country and there are many places which my camera did not visit, much to my disappointment.

I am sure most of us who took railway photographs during this period would have longed to be full time photographers, but that was never the case and it had to remain a Hobby — taking its place with all the other things that were going on in our lives.

The important thing, though, was to always carry a camera and today with the trend for miniaturisation this is comparatively easy. However, the basic precept is just as valid now as it was thirty years ago — the ever changing face of technology is with us and if you photograph the modern scene it will probably look very different in just five years time.

Quite a number of the locomotive types portrayed, have been rescued for preservation. However, in my pictures you can see the kind of condition in which they were kept when they had to earn a living, without the enormous care and attention that they now receive before each day's work.

Apart from summer evenings, when local photography was possible, the majority of pictures were taken at the weekends or during the few weeks annual holiday — we never seemed to have so many days in the Fifties as we do now! Inevitably one had a favourite location to visit, but I was lucky, as living in the Midlands I am within about 100 miles of all the mainlines of the four regions and I could easily reach them. There were special trains to consider — those run for the Railway Societies and some by British Railways, such as the Newbury Race Specials from Paddington which were always well turned out. Of course the titled trains, of which there were many, had a terrific attraction for me and the Pullman car trains on the Southern, Eastern and Western Regions were well worth the effort of a day's driving.

Without being too partisan and being allowed the luxury of photographing just one train again from the 1950s, but with modern cameras, I would go for 'The Bournemouth Belle'. It would be the 'up' train, towards the middle of May, with a stormy sky illuminated by a marvellous patch of blue sky that allows the sun to shine through the brilliant clear air. So you can set the scene — a location somewhere between Eastleigh and Basingstoke. All we need now is an immaculate rebuilt 'Merchant Navy' Class Pacific and the usual set of Pullman cars. Just remember to put a film in the camera! It is a strange thought that with a lot of organisation such a re-enactment could take place today. I should be there!

As I get older the inner need to rush around after steam specials has receded and for me just one good picture from the day's activity is all I require. Just last weekend No. 4998 *Sir Nigel Gresley* came through and the crowds of photographers, some running with enormous video cameras, dashing from one place to another was amazing to see. Perhaps I was one of the lucky ones who had it placed in front of me and I took the opportunity to the best of my ability!

For those of you who remember some of the pictures in this book I hope it will bring back happy memories and for the younger enthusiast of today who wishes to know what it was like thirty odd years ago in the steam era, all I can say is 'It was marvellous'.

R. J. Blenkinsop
1985

1 Leamington Spa Avenue Station in the middle of a winter morning, showing one of F.W. Webb's 2-4-2T of 1890 vintage with an Auto-train for Rugby. Apart from the edge of the platform there is no trace left of this scene today. The ornate supporting brackets for the station roof are worthy of study.

OAK ROOM
AND
ARCELS OFFICE

2 The track layout today has been very much simplified with a single line branching off to Stratford-on-Avon on the left. No. 6005 **King George II** approaches Hatton Station with the 11.35 Wolverhampton-Paddington and an L.M.S. coach leading, with the engine in blue livery. In the up loop a freight train is just visible hauled by ex-works R.O.D. No. 3033.

30 January 1952

3 This is a sight to thrill L.N.W.R. enthusiasts as G2a Class 0-8-0 No. 49181 accelerates away from Harbury Cement Works after the long drag up from Leamington. It was working a Nuneaton-Banbury coal train and makes a magnificent sight as the smoke hangs under the bridge on a brilliant winter morning.

2 February 1952

43210

4 You will notice the flat bottom track going down the Lickey incline to Bromsgrove as a Midland 0-6-0 No. 43210 climbs the 1 in 37 banked by an 0-6-0 tank engine in the rear. This was taken on a very dull day and just before the fireman got to work with the shovel. There was another photographer further up the hill who won a competition in 'Trains Illustrated' with his picture of this same freight train.

12 April 1952

5 I was fortunate living at Leamington as all the Big Four were close at hand. The Great Central provided my ex-L.N.E.R. motive power and in this case it is No. 60059 **Tracery** climbing past Staverton Road signal box with the up *Master Cutler* from Sheffield to Marylebone. This named train started running during the 1951 Festival of Britain.

19 April 1952

6 You may recall the British Industries Fair which was held at Castle Bromwich not far from the new National Exhibition Centre now under construction. A special was run each day from Euston in the morning and this picture is the return train in the evening. No. 45669 **Fisher** pilots class 5 No. 45064 past Cathiron with a 15-coach load, both engines being in black livery.

15 May 1952

7 Another picture on the Great Central showing the afternoon down *South Yorkshireman* approaching Staverton Road signal box. You will notice the sides of the cutting are comparatively shallow, a Board of Trade requirement when this line was built. No. 60103 **Flying Scotsman** has a rake of the latest L.N.E.R. steel coaches.

24 May 1952

THE SOUTH
YORKSHIREMAN
60103

63803

8 Turning around from the previous picture, class O1 2-8-0 No. 63803 climbs slowly up the hill towards Woodford. Both the crew are looking out on this warm summer evening and the fireman has obviously been filling up the box judging by what is coming out of the chimney!

24 May 1952

9 Passing the recently demolished L.N.E.R. signal box at Kenilworth Junction is class 5 No. 44833 with the 07.16 (S.O.) Leamington-Llandudno which started as empty stock, from Long Itchington. You will notice the signalman at his window with the staff hanging from his hand which will allow the train to proceed on the single line to Coventry. The reflecting object seen through the forward cab window is a new bucket hanging on the front of the tender.

28 June 1952

S. L. S.
SPECIAL
C SPECIAL

11 Another Stephenson Locomotive Society special which started from Cardiff and made a tour of the more obscure lines in the Welsh valleys. 0-6-0PT No. 6423 stands in Llantrisant station with two 70′ trailers, the nearest being No. 32 built in 1906.

12 July 1952

10 Engines of the 'Star' class were difficult to find in the Midlands but a number could be seen on the Swindon line in 1952. Near Wantage and on its way up to London is No. 4062 **Malmesbury Abbey** with elbow steam pipes. Taken with the sun in the west late in the evening it could not be a more difficult picture to print, but nevertheless there is movement and a blurred background to give an impression of speed.

5 July 1952

12 This picture is included for the 'environmentalists' to show what Paddington could be like but in fact seldom was so. The fireman of No. 5051 **Earl Bathurst** seems to have overdone things as the driver awaits departure from No. 2 platform. A 'Britannia' is visible in the background blowing off steam on this dull, dark morning.

4 October 1952

13 I must have travelled onto Oxford by train as this picture shows the fireman involved in uncoupling No. 34110 **66 Squadron** from a through train with engine change at Oxford. Of interest today is the leather strap for adjusting the window and the wooden slats on the carriage roof felting.

4 October 1952

34110
BRITISH RAILWAYS

THOMPSON
WAY OUT
41046
BRITISH RAILWAYS 41195
41195

14 A scene familiar to Midlanders and those who frequented Birmingham New Street station. It seems strange to look at this picture and compare it with today's smooth operation. L.M.S. compounds with Stanier chimneys Nos. 41046 and 41195 are seen complete with the Midland deep buffer beam, black at the top and vermillion below.
25 October 1952

15 The 10.15 Paddington-Wolverhampton climbs Hatton Bank on a sunny winter morning behind No. 6904 **Charfield Hall** in Great Western livery with coat of arms on the tender and No. 5960 **Saint Edmund Hall**, newly painted in black but running with a green tender.
28 February 1953

16 Another class O1 2-8-0 No. 63596 on the Great Central at Loughborough, crossing the Midland main line, the bridge being in the foreground. This is a down freight train on a dull day which I spent switching from the Great Central to the Midland, and back depending on the state of the signals!

21 March 1953

17 The preservationists of the Main Line Steam Trust will appreciate this as it shows B1 No. 61160 leaving Loughborough Great Central station with the up *South Yorkshireman*. Mainly L.N.E.R. teak stock but some already painted in that frightful cream and red.

21 March 1953

THE SOUTH
YORKSHIREMAN
61160
61160

18 Fairburn 2-6-4T No. 42160 leaves Loughborough Midland station with a Nottingham-Leicester stopping train and an L.N.W.R. 0-8-0 plods on slowly south with a freight train with the driver much exposed to the east wind. The Great Central can be seen crossing in the background.

21 March 1953

19 In fact I had no intention of taking this picture with the engine going away, but when a wall of water and spray approaches at 60 m.p.h. like this one did there is no alternative. Goring Troughs at its wettest with No. 5940 **Whitbourne Hall** on its way from Paddington to Swindon.

7 April 1953

34094
399
PASSENGERS
CROSS
LINE
SUBWAY
CHOLSEY
CHANCE

20 Cholsey and Moulsford station looking towards Reading with 'West Country' Pacific No. 34094 **Mortehoe** crossing from the down fast to the relief line as it would be taking the east avoiding line at Didcot to Oxford for a change of engines.

7 April 1953

21 At the other end of the station was Dean Goods 0-6-0 No. 2532 shunting wagons from the Wallingford branch. This engine built in 1897 was on loan to the Manchester and Milford Railway in 1906 and ran as their No. 10. It was one of the last of the class to be withdrawn in 1954 and judging by the state of the chimney was in a sorry state. The birds were building their nests high in the trees so it must have been a good summer in Coronation Year!

7 April 1953

22 This is a view looking north east at Watford station on a Sunday evening with No. 46129 **The Scottish Horse** having just pulled in with a train from Euston. Watford shed is in the background with quite a variety of motive power. The L.N.W.R. water column shows up clearly in the foreground.

19 April 1953

23 At Castle Bromwich station B1 No. 61190 runs in with a stopping train for Birmingham New Street. This was another British Industries Fair occasion, the exhibition buildings being shown on the left where skyscraper flats now abound. Notice the electric headlamps with oil lamps mounted on the top, and there is also a freight train approaching on the right.

4 May 1953

61190
61190

24 Class T9 4-4-0 No. 30285 stands at Andover station gently blowing off steam before leaving for Eastleigh where it was shedded. I am sorry the coupling rods are up inside the splashers but it is difficult to arrange for them to be visible on every occasion!

14 June 1953

25 A well-known view looking towards Waterloo with Wimbledon station in the background and yard on the right. There is a convenient footpath on the left leading down to a footbridge across the tracks. 'King Arthur' No. 30789 **Sir Guy** heads south into the high midday sun.

20 June 1953

30789
269

26 All 0-6-0 tender engines in Wimbledon Yard. On the left is No. 30694 Class 700 and in the centre is No. 30572, with No. 32524 on the right. Judging by the lack of activity in the picture it must have been the lunch hour!

20 June 1953

27 This is possibly a Waterloo-Lymington train emerging from under the shops at Wimbledon station hauled by class L12 4-4-0 No. 30434.

20 June 1953

WIMBLEDON
B
30434
30434

WH.74
WH.70
30453
30453

28 No. 30453 **King Arthur** approaches Wimbledon with a southbound train in the upper photograph and later the same
day I travelled back into London and walked over the Thames to Charing Cross Station. Two 'Schools' class
engines, both with standard chimney fortunately, are seen at the end of the platforms. No. 30935 **Sevenoaks** arrives
29 while No. 30912 **Downside** waits to leave with a train for Tunbridge Wells.

20 June 1953

CASTINGS
WEST YORKSHIRE FOUNDRIES LTD.
EUSTON
CARRIAGE SIDINGS
SLEEPING CAR
46237
46237

30 Camden Bank outside Euston early in the morning. No. 46237 **City of Bristol** has a sleeping car train from Glasgow while an 8F 2-8-0 takes empty 12-wheel sleeping cars back to Willesden for servicing. In the background an 0-6-0T shunts stock for the down *Red Rose* Euston-Liverpool. Note the L.N.W.R. signal, the steam from the tender coal pusher, and the remains of streamlining on the top of the smokebox.

25 July 1953

31 This is undoubtedly one of my best pictures, looking down Camden Bank into Euston which is just out of the picture on the left. No. 45721 **Impregnable** has a push at the back of the train by an 0-6-0T and above the **Jubilee** are two more 0-6-0T's with a rebuilt **Royal Scot** attached to its train which was to form the down 'Irish Mail'. You will also notice the enormous carriage shed in the background and an '8F' on the left.

25 July 1953

32 No medals for guessing this location but I do recommend that you study carefully the external condition of No. 60028 **Walter K. Whigham** as it stands at Kings Cross awaiting departure with the down *Elizabethan.* Notice the burnished buffers, cylinder covers and all the valve gear.

27 July 1953

33 The down **Royal Scot** emerges from Primrose Hill tunnel and into the closed South Hampstead station. The Great Central out of Marylebone crosses just above the tunnel mouth and No. 46224 **Princess Alexandra** is in green livery with the latest headboard. Those 'Virol' advertisements must have adorned railway property far and wide.

27 July 1953

ROYAL SCOT
46224
VIROL

34 I spent a day at Rhyl on the main Chester-Holyhead line and this picture shows a train entering the station from Llandudno comprising L.N.E.R. stock with class 5's No. 44714 and 45195. Of interest is the large bell on the side of the L.N.W.R. signal box above the steam

35 Standing beside the water tank at Peterborough is K3 2-6-0 No. 61821 having its tender replenished. There appears to be a crew of three but perhaps the man pointing his hand came off another engine. The design of the water column is interesting with its operating lever.

61821
BRITISH RAILWAYS

No 990
G N R
990

36

The reason for going to Peterborough was to see a special train from London to Doncaster for the Centenary Celebrations of the works. For this occasion two Ivatt 'Atlantics' were removed from York museum and subsequently worked a number of special trains before going back to the museum. Both these pictures were taken from the first road bridge to the north of Peterborough station which required dashing across the busy road and winding on at the same time. It was again one of those lucky occasions when the sun was not wanted as it was immediately behind the train and it would have ruined the picture had it been shining. Peterborough Cathedral shows up clearly on the skyline with the brickwork chimneys in the distance.

20 September 1953

37

41163
41163

38 Standing in the up platform at Coventry Station with a local train to Rugby is compound 4-4-0 No. 41163. This picture gives a good impression of the old station with its two platforms and fast lines in the middle. Note the ornate cast brackets for supporting the roof.
24 September 1953

39 The first express out of Worcester in the morning carried the 260 reporting numbers and it is shown here entering Paddington under steam, probably having been held by signals at the entrance to the station. No. 5063 **Earl Baldwin** is the 'Castle' class engine hauling this train with more 'Virol' advertising in the background.
26 September 1953

58071
6807
M 20450 M

40 A delightful sight at Gloucester standing between the two stations, 0-4-4T No. 58071 built by Neilson & Co. in 1883 for the Midland Railway. It retained the Johnson Class C round top boiler and was one of ten fitted with condensing equipment for working the Metropolitan widened lines. Note the elegant shape of this engine with its Salter safety valves as it awaits duty on a foggy November morning.
18 November 1953

41 A ray of sunshine slants through the fog to show a busy scene just outside Gloucester. The Great Western shed is on the right with an 0-4-2T on the turntable. On the left 0-6-0 class 4F No. 44567 shunts wagons in the sidings outside Eastgate station. Jinty 0-6-0 No. 47607 is on the left and the variety of signals is a pleasure to study.
18 November 1953

6918
NO ADMITTANCE

42 This photograph shows the tramway crossing at the back of the Great Western Gloucester shed. It is a crossing widely known for its disruption to road traffic and the fantastic variety of motive power from many railways which have used the Midland and Great Western lines. No. 6918 **Sandon Hall** still in Great Western livery has just left Gloucester Central station and is passing the crossing keeper's hut.

18 November 1953

43 I have included this picture as it shows wrong line working on a Sunday between Leamington and Warwick. The train is the 11.10 Paddington-Birkenhead hauled by No. 6018 **King Henry VI** which was the first of the 'Kings' I saw to be repainted green after their few years in blue which somehow did not suit them and looked terrible when dirty.

14 February 1954

44

It makes one wonder at which end of the engine was fitted the water scoop as No. 4977 **Watcombe Hall** passes over Lapworth troughs with an up freight. I assume it is the A.T.C. equipment creating all the spray the like of which I have never seen before or since.

20 February 1954

45

The exhaust from No. 6844 **Penhydd Grange** blackens the sky as it approaches Stratton St. Margaret with a down freight. Shedded at Llanelly it was probably on its way back to South Wales.

6 March 1954

6844

45310

46 Both drivers are leaning out of their cabs as they emerge from Kilsby Tunnel into the bright sunshine with a Wolverhampton-Euston express. There was a P.W. slack in operation at that time so class 5 No. 45310 and No. 45641 **Sandwich** are moving slowly over the newly laid and unballasted track.

27 March 1954

47 This was quite a lucky shot just to the south of Rugby Great Central station, showing A3 'Pacific' No. 60050 **Persimmon** accelerating down the hill with the up *Master Cutler*. K3 2-6-0 No. 61980 has a freight from Woodford on its way north with the steam from both engines showing up clearly on a lovely spring morning.

27 March 1954

48 My old friend the 11.35 Wolverhampton-Paddington in Harbury cutting hauled by No. 6016 **King Edward V**. This is a somewhat unusual view as the trees and bushes had just been cut down on the west embankment. By the end of the year they had grown up again and have never been cut since, making photography impossible.

27 March 1954

49 The down *Pembroke Coast Express* approaches Twyford with tall chimneyed No. 5089 **Westminster Abbey.** The old type of aluminium headboard is carried and the train comprises a set of Great Western coaches.

19 April 1954

PASSENGERS ARE REQUESTED TO CROSS THE LINE BY THE BRIDGE
30862
395

50

To achieve a reasonable quota of pictures I had to cheat a little and use the through trains which had Southern motive power southwards from Oxford. No. 30862 **Lord Collingwood** has just come off Reading West curve and enters the station where it will stop. The new British Rail coaches contrast with the rather 'worse for wear' Great Western signals.

19 April 1954

51

Totally unexpected, this excursion has just passed through Coventry station on its way to Birmingham. B1 No. 61138 looks spick and span with its set of North Eastern stock and this view is of interest now that the scene has changed so much due to electrification. They must have had fun taking down that tall L.N.W.R. signal!

27 May 1954

5972
BRITISH RAILWAYS

52 This is Hatton Bank but not a view often seen in pictures. My open Morris Minor stands on the bridge so it must have been a warm evening as No. 5972 **Olton Hall** travels onto Warwick with a semi-fast from Birmingham Snow Hill. I think the engine came from Tyseley and is in black livery.

4 June 1954

53 The next five pictures were all taken at Worting junction just to the south of Basingstoke where the line to Salisbury and the west leaves the main line from Waterloo to Bournemouth. No. 30457 **Sir Bedivere** comes round the curve and under the flyover on its way to Salisbury. What an exciting change it was to see these handsome engines and green coaches.

7 June 1954

35003
35003
4 | 5

54 On its way up to London from Exeter is No. 35003 **Royal Mail** of the 'Merchant Navy' class.

7 June 1954

55 And now coming down from Waterloo past the flyover and onto Micheldever is 'West Country' Pacific No. 34020 **Seaton**. These were always difficult engines to photograph unless the sun was shining on to the smokebox as the smoke deflectors and top cowling invariably caused nasty shadows on the front.

7 June 1954

30450
30450

56 This one is taken from the flyover looking west as an up train from Salisbury approaches behind No. 30450 **Sir Kay**.

7 June 1954

57 The final picture at Battledown flyover shows an up train from Bournemouth behind 'West Country' Pacific No. 34093 **Saunton**. You notice the magnificent sky and the banks of cloud coming up from the south west. As luck would have it the clouds stayed on either side of the sun for most of the morning.

7 June 1954

58 On my way home I stopped at the south end of Cholsey cutting to see yet another through train from the north which picked up a Southern engine at Oxford. In this case it is No. 30742 **Camelot**.
7 June 1954

59 Next day was stormy but justified this picture of 2-6-2T No. 5161 coming up the last mile to Leamington Spa with a pick up freight which would probably go straight into the goods yard. The engine was shedded at Leamington and is still in Great Western livery.
8 June 1954

5161
G W R

60 Waterloo station on a dull day with 'Merchant Navy' class Pacific No. 35011 **General Steam Navigation** awaiting the "right away" with the down *Bournemouth Belle*. Of all the titled trains the *Belle* always gave me the greatest thrill as it swept by with its set of Pullman cars and attendants in their smart uniform.

31 July 1954

61 London commuters may recognise this location with West Hampstead tube station on the right. No. 60054 **Prince of Wales** is about to start the final part of its journey through the tunnels down the hill to Marylebone with the up *Master Cutler*. The engine was in blue livery.

2 August 1954

62 Hatfield was not far out of my way when returning home to Leamington and I stopped at the sweeping curve to the south of the station. No. 60149 **Amadis** has an up express while V2 No. 60867 passes by in the opposite direction. Although very much a dull weather photograph with two dirty engines, it is still a view very much part of the everyday running of a railway.

[illegible] August 1954

63 'Britannia' Pacific No. 70045 was unnamed when this picture was taken as it passes the Point of Air on the North Wales main line with the 15-coach down 'Irish Mail'. You will notice the coal is well down in the tender, and just visible shunting on the right is Lancashire and Yorkshire 0-6-0 No. 52356.

16 August 1954

70045
70045

64 Here we have another atmospheric offering as 2-8-2T No. 7212 toils up Campden Bank with a loaded up coal train shortly after leaving Honeybourne. I was on my way to Devon for a quick weekend of photography and from the car noticed the smoke in the distance so stopped and quickly found a suitable spot by the roadside.

30 August 1954

65 No. 45738 **Samson** passes Radford Brewery with a diverted Sunday Wolverhampton-Euston express. The Leamington-Rugby line is now removed here and Radford Brewery demolished and replaced by the offices of the East Midlands Electricity Board. The Thornley family who owned the brewery were great users of the famous Foden steam wagon for delivery of their beer but before my time I'm afraid.

45738

66 I have included this picture as it shows an old signalman friend of mine at Southam Road and Harbury signal box together with his bicycle which was the standard form of local transport in those days. A rather grubby 'King' No. 6014 **King Henry VII** passes by with an up express from Wolverhampton. The scene today is very different with just two main lines, and no sidings, station or signal box.

20 February 1955

67 A classic train which always had 6P or 7P motive power. It left Leamington Avenue station at 07.55 as a stopping train to Birmingham where it became an express on to Liverpool. The return working arrived in Leamington in the early hours of the morning. This picture shows it rounding the curve at Kenilworth Junction, taking the Berkswell line. 'Royal Scot' No. 46151 **The Royal Horse Guardsman** produces a realistic shadow on the embankment as its exhaust billows forth on a cold morning.

28 March 1955

KENILWORTH JUNCTION
46151

68 Southern engines were not fitted with water pick-up equipment as the Southern Railway did not really require water troughs due to lack of long non-stop runs. 'Lord Nelson' class 4-6-0 No. 30861 **Lorn Anson** passes over Goring troughs on its way from Reading to Oxford. **2 April 1955**

69 Ian Allan Ltd. ran a special from Paddington-Bristol-Birmingham and back to London and No. 7017 **G.J. Churchward** hauled the train to Bristol and Birmingham. It was routed on the Midland line up the Lickey incline. This picture shows the famous 0-10-0 No. 58100 banking the train with the 'Coronation' beaver tail observation car at the rear of the train.

16 April 1955

180
CITY OF BRISTOL
46237

70 A Stanier 'Pacific' was loaned to the Western Region for trials and comparison with the 'King' class engines of the Great Western. The engine selected was No. 46237 **City of Bristol** and, before working the trials proper with Dynamometer car on the Plymouth run, it spent a few days coming down to Birmingham. Here it is passing Hatton North Junction with the 09.10 Paddington-Birkenhead.

27 April 1955

71 The driver is seated peering through the cab front window of No. 6005 **King George II** as it descends Hatton Bank with the 15.00 Birmingham Snow Hill-Paddington. The tender top is interesting, covered with coal or a mixture of coal and water slopping around in the back.

18 June 1955

30864
255
TRAINS
DANGER

72 The next six pictures were all taken on my way back from my honeymoon in Jersey. The boat express from Southampton to Waterloo was hauled by No. 30864 **Sir Martin Frobisher** and is shown in the entrance to the customs house where the rest of the train was stabled. Class USA 0-6-0T No. 30063 stands on the left of the picture.

9 July 1955

73 **Sir Martin Frobisher** pulls slowly out of Southampton Docks across the main road and on its way to Waterloo through the terminus station. The beautiful face of the South Western Hotel shows up clearly in the bright afternoon sunshine.

9 July 1955

74 Both these pictures were taken at speed from the carriage window. Eastleigh shed has a varied selection of motive power on display, and I think it is best for you to test your powers of recognition. Numbers are hardly legible but the two 'King Arthurs' on the right are No. 30767 **Sir Valence** and No. 30764 **Sir Gawain**.

9 July 1955

75 I cannot make up my mind if an M7 0-4-4T is male or female. Someone has certainly made up their mind as "Pat" appears quite clearly on the front of the smoke box. No. 30378 is running into Eastleigh with a stopping train from Winchester.

9 July 1955

852
PAT
30378

31630
31630

76 We had to change at Basingstoke for the journey onto Leamington and there was time to take a few pictures. Running into Basingstoke from the Reading line is class U 2-6-0 No. 31630 with a set of vintage coaches.

9 July 1955

77 Turning round from the previous picture my camera was pointing at 'Remembrance' class 4-6-0 No. 32330 **Cudworth.** This engine, one of a small class of seven was a rebuild from L. Billington's 'L' class 4-6-4T's built between 1914 and 1922.

9 July 1955

78 No. 45544, an unnamed member of the 'Patriot' class, is seen backing out of Liverpool Lime Street station past the usual crowd of engine spotters who now have electrics and diesels to fill their notebooks.

16 July 1955

79 Looking very smart in its new coat of paint, class C14 4-4-2T No. 67442 stands in Neston North station with a train for Wrexham. Judging by the number of heads appearing from the window there were quite a number of people travelling on this Sunday morning. The station architecture is worthy of note together with the water column.

67442
67442

80 Two studies of class N5 0-6-2T in the Wirral. No. 69362 leaves Burton Point with a train for Wrexham.

3 September 1955

81 and in the lower photograph No. 6926[illegible] starts away from Chester Northgat[illegible] station now closed and demolished.

3 September 195[illegible]

69267

82 By Chester golf course in the evening the up 'Irish Mail' approaches behind 'Britannia' Pacific No. 70046 while 'Class 5' No. 45002 accelerates out of the cutting on its way along the North Wales coast.

3 September 1955

83 There seems to be quite a mixture here in Great Western territory at Aynho water troughs, where a culvert bridge is being replaced over the troughs. It was single line Sunday working and the two 'Directors' class D11 4-4-0 No. 62666 **Zeebrugge** and No. 62667 **Somme** are working a special train to the Farnborough Air display.

11 September 1955

E 1596 E
62667
62666

84 The scene is Shrewsbury Great Western shed with visiting class T9 4-4-0 No. 30304 being prepared for the Tallylyn Railway Preservation Society special which it worked to Towyn together with 'Dukedog' 4-4-0 No. 9027 coming on at Welshpool.
24 September 1955

85 A class G2 0-8-0 climbs up to Runcorn bridge with an up freight on the main Liverpool-London line, now of course electrified. I seem to recall that this picture was taken from the transporter bridge on one of my numerous journeys across while hoping a train would appear.
15 October 1955

COSSAGES

86 The weather was miserable on every Saturday while the "King crisis" was on; you may remember they were all withdrawn after faults appearing in bogies and main frames. In this picture No. 46254 **City of Stoke-on-Trent** leaves Leamington with the 09.10 Paddington-Birkenhead. For the record there were four Stanier 'Pacifics' on loan, namely Nos. 46207/10 and 46254/7.

28 January 1956

87 A real example of push and pull with 0-4-2T No. 1416 in the middle heading into Chester and crossing the River Dee with the race course on the right. A freight train is disappearing on the up main line, and coming this way on the down relief a Great Western 'Hall' is towing a 'Mogul' and a British Rail standard class 5 No. 73099.

10 March 1956

44829

89 The 'King' class had now all returned back in service and No. 6007 **King William III** is shown on the climb up to Fenny Compton with the 15.00 Birmingham-Paddington. The exact location is just short of the earth works which were begun and of course never finished to take the Great Western line through Southam.

2 April 1956

88 The Royal train leaves Coventry after H.M. The Queen had laid the foundation stone for the new Coventry Cathedral. Class 5's No. 44829 and No. 44833 pass Humber Road signal box with the spire of the old Cathedral visible on the right. The first coach was part of the L.N.W.R. 'Royal train' built at Wolverton and the remaining coaches are the 1941 stock.

23 March 1956

90 The two sides of Catesby tunnel are shown in these two pictures. Class O1 2-8-0 No. 63887 has a down freight from Woodford about 1 mile north of the tunnel which can just be seen above the platelayer's hut.

12 May 1956

91 Believed to be the first time a Pullman car train had worked out of Marylebone, the 'Pennine Pullman' called at Sheffield over the Pennines and back up the East Coast main line. No. 60014 **Silver Link** is about to enter Catesby tunnel travelling at just under 60 m.p.h. having shut off steam for a P.W. slack.

12 May 1956

Ian Allan
Trains
Illustrated
EXCURSION
60014
60014
A-4

92 Following the 'Pennine Pullman' was a service train headed by No. 60063 **Isinglass** and about to pass through Charwelton station. Many of the Great Central stations were sited between the running lines and you will notice here the lines opening out to allow room for the platform.

12 May 1956

93 The 'Cornishman' ran from Wolverhampton to Penzance in both directions. The train from Wolverhampton is shown here coming off the west triangle at Hatton behind No. 5070 **Sir Daniel Gooch.** The line to the right goes to Hatton station and on to Leamington. The west triangle has now been singled as has the branch as far as Bearley.

15 May 1956

THE CORNISHMAN
825

94

The fastest train in the book ends this volume in the middle of 1956. I am sure No. 7030 **Cranbrook Castle** was travelling at over 70 m.p.h. as it passes Ruscombe halfway between Maidenhead and Reading. The train is of course the down 'Bristolian'.

22 May 1956

Echoes of the
BIG FOUR

R. J. BLENKINSOP

1 No. 6011 **King James I** with the 17.10 Paddington-Wolverhampton is travelling over 70 m.p.h. as it descends the bank between Fosse Box and Leamington Spa. Taken on a fine summer evening with sun behind the camera, the driver can be seen standing on the far side of the cab and the fireman sitting and waiting for the down distant signal to come into view.

25 June 1956

2 The cutting here has been filled in and little trace of a railway can be seen. Through the bridge is the estuary of the River Dee with the Welsh mountains behind. 2-6-2T No. 40101 climbs away from Neston South with a local train from West Kirby to Hooton.

31 July 1956

3 Newton Abbot had a stud of 'Manor' class engines to assist trains over the gradients between Aller Junction and Plymouth. The up *Cornish Riviera Limited* is shown entering the tunnel at the summit of Dainton Bank hauled by No. 6010 **King Charles I** and assisted by No. 7813 **Freshford Manor**. Engines employed in banking freight trains from both directions would use the double slip points to return to Newton Abbot or Totnes.

4 August 1956

4 An up morning express from Plymouth is seen passing through Plympton station at the foot of Hemerdon Bank hauled by No. 7814 **Fringford Manor** and No. 6906 **Chicheley Hall**. Gas lamps are visible on the platform as well as the signal box which as a vantage point, gave a wonderful view of the trains.

6 August 1956

5 In the afternoon storms swept in from the Atlantic and the visibility from the top of Hemerdon was poor to say the least. However, I could not resist this one as it was slipping all the way up the 1 in 42 and only near the top did it get its feet properly. Consisting mainly of North Eastern stock No. 34032 **Camelford** of the 'West Country' class is in charge.

6 August 1956

6 As a final gesture to the poor weather I stopped in a lane near the foot of Hemerdon Bank and as luck would have it the sun shone through the clouds at the right moment. No. 7809 **Childrey Manor** and No. 7036 **Taunton Castle** pass with an up express.

6 August 1956

7 Cornwood station the following morning, a very rural scene with 0-6-0 PT No. 3675 passing through with a pick-up freight from Plymouth. The station flowers are in full bloom and passengers are waiting the arrival of a stopping train to Totnes.

7 August 1956

8 Whilst the summer services produced more trains to photograph the external appearance of the engines suffered from lack of cleaning as is shown in this picture of the down *Cornishman* near Bishopsteignton running along the River Teign. No. 6873 **Caradoc Grange** and No. 5071 **Spitfire** have a light load of eight coaches.

8 August 1956

9 Super power indeed, ten coaches and two 'Castle' class locomotives running along the seawall just before turning north at Dawlish Warren station. No. 7029 **Clun Castle** and No. 5089 **Westminster Abbey** provide a fine sight on an up train as storm clouds gather and the sun slips behind an overcast sky.

9 August 1956

7029

34049

10 'Battle of Britain' class Pacific No. 34049 **Anti-Aircraft Command** joins the Western region main line at Cowley Bridge Junction with a stopping train from the Barnstaple line or perhaps it originated from Plymouth via Okehampton. The engine is carrying a Salisbury shed plate.

9 August 1956

11 A picture taken at Shrewsbury shed just before the Wainwright 4-4-0 Class D No. 31075 set out for Towyn with a Talyllyn Railway Preservation Society Special. Inspector Holland is standing on the left with the shed foreman on the right and No. 4942 **Maindy Hall** is partly hidden in the background.

22 September 1956

WELSHPOOL
JUNCTION FOR
SHREWSBURY, STAFFORD,
BIRMINGHAM, LONDON
TRPS SPECIAL
31075
31075
SPL
14

12 The T.R.P.S. special has now arrived at Welshpool and both engines take on water. Dean Goods 0-6-0 No. 2538 is the train engine but perhaps the standard Harris Tweed jackets and grey flannel trousers of the enthusiasts are of more interest!

22 September 1956

13 A 'King' in full flight starting the climb to Hatton alongside Warwick gas works. On the up line the through train from Birkenhead to Ramsgate is passing as No. 6007 **King William III** has just emerged from under the Cape road bridge with the 09.10 Paddington-Birkenhead.

3 November 1956

PNX 759

14 I have included this picture to show the old type of refuse collection vehicle with crew taking a rest from emptying the bins. Probably it would qualify as a vintage commercial vehicle today. The local train will have come from Birmingham behind 2-6-2T No. 4112 and has just crossed the A41 out of Warwick. **16 February 1957**

15 A cloudless winter's day with a deserted summit of Hatton Bank. 2-6-2T No. 6116 heads a local train for Leamington Spa and No. 5035 **Coity Castle** with leaking inside cylinder glands comes up the gradient with the down *Cambrian Coast Express.*

16 February 1957

25
40
6011

16

This train carried through carriages from Birkenhead to Paddington. No. 6011 **King James I** came on at Wolverhampton with a departure time of 11.35 and it is shown accelerating out of Leamington Spa and past the G.W. shed on its way to London, stopping at Banbury.

2 March 1957

17

With the coming and going of mainline trains from the platforms off the picture on the right, the suburban trains were hardly noticed as they crept away from King's Cross and into the first tunnel. Class N2 0-6-2T No. 69577 is fitted with condensing apparatus and shorter chimney for working on the Metropolitan lines to Moorgate.

23 March 1957

61200
61200

18 The next seven pictures were all taken one afternoon that I spent alongside the G. N. mainline just north of Welwyn Garden City. Class B1 No. 61200 is leaving the station with a semi-fast train most likely for Cambridge.
23 March 1957

19 Coming up to London from the north is Class A1 No. 60139 **Sea Eagle** returning to its home shed at King's Cross. Lack of exhaust on this warm spring day was made up by the sight of this beautifully clean engine as soon as it was seen emerging from Welwyn tunnel in the distance.
23 March 1957

20 A black and white picture can hardly do justice to the sight and sound of this train which was audible long before it came into view. The three cylinder A3 Pacifics are unique as a musical experience and No. 60044 **Melton** is living up to its reputation with a down Leeds express—just look at that exhaust!

23 March 1957

21 In the late afternoon a fully fitted freight used to appear behind an express engine on its way north. This seemed to be a regular working and with rattling four wheel wagons swaying from side to side behind its enormous tender No. 60007 **Sir Nigel Gresley** is definitely out of place on this type of work. Note the state of the track on the down slow line.

23 March 1957

60007

90559

22 It was my intention to only photograph No. 60800 **Green Arrow** which was slowing for signals on the up fast line. However a very clean 'Austerity' 2-8-0 No. 90559 was coming along the down slow line, so with a dash along the embankment both trains were included, but as you can see the shutter was fired a fraction of a second too late.

23 March 1957

23 A pair of articulated coaches are behind the tender of Class A1 No. 60120 **Kittiwake** as it thunders through Welwyn Garden City station with a down express. Of more interest is the Class L1 2-6-4T arriving at the platform with water pouring out of the tanks as the brakes are applied too quickly.

23 March 1957

24 Class B17 No. 61671 **Royal Sovereign** enters Welwyn Garden City with a train from Cambridge. This locomotive was often used for hauling the Royal Train and was shedded at Cambridge where it was kept in fine mechanical and external condition.

23 March 1957

25 Southcote Junction looking towards Reading with the mainline to Exeter and Plymouth disappearing out of the picture on the left. 'King Arthur' class No. 30780 **Sir Persant** passes by with a through train from the north to Bournemouth Central.

13 April 1957

30780

26

It was unusual to see such a clean locomotive on an up freight train but I think No. 46152 **The King's Dragoon Guardsman** was running in after a major overhaul at Crewe works. The location is looking down from the bridge which carries the A53 over the mainline at Whitmore station.

20 April 1957

27

This day was spent looking at L.M.S. engines starting at Whitmore and ending at Edge Hill. It was a lovely sunny day and Class 5 No. 45060 is seen leaving Chester with an excursion along the North Wales coast. These locomotives, I feel, look photogenic from whatever angle the picture is taken.

20 April 1957

W561
45060
45060

45527

28

The long deep cutting with its arched supports gives away the location of this picture, it must surely be Liverpool Lime Street, but no signs of electric traction here. No. 45527 **Southport** emerges into the sunlight with a train from Euston, carefully observed by the engine number collectors on the platform.

20 April 1957

29

The other end of the cutting merges into Edge Hill station and this is an evening train coming up the steep gradient behind No. 46114 **Coldstream Guardsman.**

20 April 1957

44914
44914

30 As I have mentioned before in my Western Region Albums, Cup Final day was always important in the photographic diary as many special trains were often required and without the vandalism which takes place today. Just to the south of Rugby, at Hillmorton, the early Wolverhampton to Euston train has been strengthened and has the addition of a Class 5 at the front. No. 44914 acts as pilot to No. 45643 **Rodney** as they head south with fifteen coaches.

4 May 1957

31 Some minutes later two specials pass on their way to Wembley. On the left is B.R. Class 5 No. 73092 which will travel via Northampton, and on the right 'Patriot' class No. 45528.

4 May 1957

41285
41285

32 On the left you can see the wireless masts of the B.B.C. Rugby transmitters and 2-6-2T No. 41285 is propelling an auto train stopping at all stations to Northampton.

4 May 1957

33 A picture full of interest at the south end of Rugby station. In the background can be seen the station and also the bridge carrying the Great Central mainline to Marylebone. G2 0-8-0 No. 49447 has used the flyover on the Northampton line to cross over to the south side of the station and it is so dirty that you cannot see the steam dome cover. With another Cup Final special Class 5 No. 45257 heads for Wembley.

4 May 1957

34 At the junction near Rugby No. 7 Box the line from the north swings round to join the original line from London to Birmingham. Today there is a flyover to avoid the situation shown in this picture where the Class 5 on the left has to be held at signals to allow a train from the north to have priority. The Class 5 is on a special from Birmingham to Wembley and No. 46122 **Royal Ulster Rifleman** will no doubt be getting there first.

4 May 1957

35 Another lucky shot, but I had been hoping for a clean engine on the down *Royal Scot* as was usually the case. No. 46241 **City of Edinburgh** passes No. 7 Box and on the left is G2 0-8-0 No. 49342 slowly heading for Rugby station and the south.

4 May 1957

49342
46241

36

From time to time I have difficulty in remembering the exact location of some of these pictures as they were taken up to nearly twenty years ago. The G2 0-8-0 is making good progress on its way north to Shrewsbury and I think the bridge in the background carries the A472 down to Pontypool, and yes I have failed to record the number of the engine.

18 May 1957

37

Now we are at Pontypool Road and No. 5004 **Llanstephan Castle** passes a goods yard with a through Manchester to Cardiff train. Note the loaded coal wagons on the right of the picture, a scene which has drastically changed today.

18 May 1957

318

38 I seem to be very lucky in the number of pictures showing two trains passing each other and here is another example taken in the eastern outskirts of Newport. No. 5918 **Walton Hall** heads for the Welsh capital with a train from Bristol and approaching is W.D. 'Austerity' No. 90323.
18 May 1957

39 Nearing the end of its working life, as it was withdrawn in 1960, the Brown-Boveri Gas Turbine passes Twyford Box with the early morning Bristol-Paddington express. It had a short life, originally ordered by the Great Western Railway, and put to work in early B.R. days during the spring of 1950. I can recall many visits to Sonning Cutting near Reading and hoping for a 'Castle' on the 'Whistlers' working but even so sense prevailed and I built up quite a good set of pictures.

11 June 1957

45613

40 Now we have a change to the north of England as a few hours were spent by the main line at Shap after a holiday in the Lake District. My only visit to Shap and alas I did not have time to get to know all the positions for the best photographs. Late one evening a freight came up with No. 45613 **Kenya** piloting what looks like an 8F 2-8-0 and a banker at the rear of the train.

26 June 1957

41 The following day on our way south we stopped at Hest Bank station to see the up *Caledonian* pass through behind No. 46242 **City of Glasgow**, the locomotive badly damaged in the Harrow and Wealdstone accident on 8 October 1952. Note the water troughs beyond the footbridge.

27 June 1957

45235

42 Farther south at Chester the same day I saw the evening rush hour out of the city and this included a nice clean Class 5 No. 45235 heading westwards into the sun and down the coast to Llandudno.

27 June 1957

43 Weaver Junction where the Liverpool line leaves the mainline to the north and No. 46253 **City of St. Albans** hauling 15 coaches carries the headboard of the up *Red Rose*, which was the late afternoon express from Liverpool to London.

28 June 1957

46228

44 In the evening I motored up to Moore Troughs just south of Warrington and saw a number of expresses from the north. No. 46228 **Duchess of Rutland** gives the track a further soaking as it picks up water—the water tower being clearly visible in the background.

28 June 1957

45 Following the Perth express shown in the previous picture, the next train came from Blackpool and makes a fine sight in the low sunshine behind No. 45653 **Barham.** At this period the coaches were painted maroon but, as in this photograph, there were examples of the old red and cream livery still in use.

28 June 1957

46 About a month later I spent a useful afternoon near the mouth of Welwyn North Tunnel and had the opportunity of seeing and hearing some of the eastern engines. A pullman car train is seen approaching behind an unrecorded engine and in the foreground after much running I managed to get No. 60853 on a down express.

20 July 1957

47 For those of you who know the three cylinder Gresley engines well, the next two pictures do not really require any comment. V2 No. 60975 comes out of the tunnel with an express from King's Cross.

20 July 1957

48 A little nearer the tunnel mouth this time and A3 Pacific No. 60056 **Centenary** bursts into the sunshine heading for Leeds.

20 July 1957

49 The object above the cab of No. 60029 **Woodcock** is a short signal post and I had hopes that the engine would cover it as it went by, but you invariably find out when it is too late. The train is from Scotland nearing the end of its journey.

20 July 1957

50

Class L1 2-6-4T No. 67745 slows down for a stop at Welwyn North station and is framed by a convenient set of signals.

20 July 1957

51

This view is looking north west at Welwyn viaduct and shows an empty stock working headed by an A3 Pacific. It gives a good idea of the size of these magnificent brick built structures which are a continual source of interest.

20 July 1957

52 A photograph which always will give me a lot of pleasure and I call it my 'Tiger Moth' picture as an aeroplane of this type is visible just above the fourth coach. We are looking south east and the A3 Pacific, with firehole door open, heads for King's Cross.

20 July 1957

53 No. 46242 **City of Glasgow** and shedded at Camden is still on the *Caledonian* working shown previously in picture No. 41. In this case it has just passed through Bulkington station with the evening northbound train.

1 August 1957

54 I used to spend frequent weekends with my wife's parents who live near Chester and this has always given easy access to the Midland Region of B.R. This cloudless day started at Chester with a view looking over the road bridge by the station. Note the smoke coming from an engine leaving the Northgate station and no less than eight other locomotives visible as a train leaves for Shrewsbury behind No. 7922 **Salford Hall**.

3 August 1957

55 After a quick drive down to Whitmore on the North Western mainline, I was in time to see the down *Comet* come round the curve behind No. 45644 **Howe** on its way from Euston to Manchester.

3 August 1957

W456
45546

56 In the opposite direction on a special working was unrebuilt 'Patriot' No. 45546 **Fleetwood** with the fireman leaning out of the cab after some work with the shovel, the effect of which is beginning to show from the chimney.

3 August 1957

57 One of the fascinations of watching trains on a summer Saturday in the steam era was the sheer quantity of express working that passed by. On this stretch of the line there was practically a train audible all the time and the four tracks were in constant use. No. 45545 **Planet** comes past with a Birmingham to Glasgow express.

3 August 1957

W74

58 You may be surprised to see yet another shot of No. 46242 **City of Glasgow** but at that time it was kept in good external condition, painted in maroon livery, and was out most days of the week. Here it has just passed under the A53 with Whitmore station in the background and is hurrying south with a Glasgow express.

3 August 1957

59 Now I have moved half a mile north to the water troughs and these two Class 5 locomotives appear to be having a race on their way towards Stafford. On the left is No. 45287 and on the right, a real coincidence No. 45288. I bet that has never happened before!

3 August 1957

60 This is the best angle for the 'Royal Scot' class to be photographed and No. 46111 **Royal Fusilier** has just breasted the summit on the climb out of Crewe with a heavy load of fifteen bogies.

3 August 1957

61 In the evening I called in to a spot overlooking the racecourse at Chester with, in the foreground, the bridge carrying the North Wales line over the River Dee. At this time there were not too many Compounds still working but I was lucky to see No. 41119 in a respectable condition, passing the racecourse set up for the Agricultural show.

3 August 1957

41110

62 And so back to base at Leamington Spa with the 09.10 Paddington to Birkenhead passing underneath the Grand Union Canal which forms the dip between the station and the start of Hatton Bank. Usually worked by a 'King' class engine on this particular day No. 7027 **Thornbury Castle** deputises, and it is good to know that it may run again as it is now at the Birmingham Railway Museum.

17 August 1957

63 Stanier 0-4-4T No. 41902 passes Radford Brewery with the 07.30 Leamington Spa to Napton and Stockton. This train I have photographed many times as we met every morning while on my way to work in Coventry.

21 August 1957

41902

64 At the end of August I spent part of my summer holiday in Devon and I am including three pictures all taken on the same day. No. 1016 **County of Hants** in immaculate condition has just passed Dawlish station with the 08.00 Plymouth to Crewe, changing engines at Shrewsbury. The double chimney does not look too bad in this shot but generally I feel the Counties were not improved when they lost their single chimney.

29 August 1957

65 Starting its journey up the estuary of the River Teign No. 6385 has a stopping train from Exeter to Newton Abbot and has just left Teignmouth. Note the polished safety valve bonnet and the clean appearance even in the height of the summer services.

29 August 1957

66 It is not all that easy to obtain pictures of trains with a clear seascape in the background but this is one such spot between Dawlish and Teignmouth where the line runs by the sea and one can climb up the embankment. No. 34061 **73 Squadron** has a local train stopping at all stations between Plymouth and Exeter.

29 August 1957

67 You may think that this is deep in Southern territory but it is not—in fact the main Great Western line from Paddington to Plymouth is shown just to the south of Reading West station. Two 'King Arthur' class engines are passing with trains between Reading and Basingstoke. No. 30771 **Sir Sagramore** is in the foreground and No. 30785 **Sir Mador de la Porte** approaches from the south.

7 September 1957

30785
30771

30783

68

Another Southern engine leaving Reading West station with a through train from the north to the holiday resorts on the south coast. In this case it is No. 30783 **Sir Gillemere** with quite a mixture of coaching stock from the Midland and Eastern Regions.

7 September 1957

69

On the way home I spent an hour at Goring watching the engines picking up water from the troughs. No. 2845 has a set of empty stock both Midland and Western Region and its A.T.C. shoe can be seen just skimming above the water in the trough.

7 September 1957

70 In the depths of winter No. 60008 **Dwight D. Eisenhower** has just passed through Welwyn Garden City with the down *Flying Scotsman* and starts a downhill run for the next few miles.

11 January 1958

71 No. 34054 **Lord Beaverbrook** approaches Vauxhall station with a lightly loaded train and a Southern E.M.U. is about to overtake.

12 April 1958

32487
791

72

The driver of this engine looks incredibly small leaning out of the cab as No. 32487, an 0-6-2T introduced in 1897, is on its way to the carriage sidings with a load of empty stock from Waterloo station.

12 April 1958

73

Quite a cheerful smile from the fireman of this pannier tank leaving the goods loop to the south of Ruabon station. The engine, No. 7403, has a guard's van in tow and will be at once turning right at Llangollen Junction where the line to Dolgellau leaves the Shrewsbury to Chester mainline.

26 April 1958

74

Part of the platform at Bulkington station is made of concrete slabs but there is also a considerable area of wooden planking. My faithful Morris Minor stands in the road and of course the scene today has changed beyond all recognition. Class 8F No. 48194 in the late afternoon sunshine takes a load of empty coal wagons down towards Rugby.

6 May 1958

75

1958 was to be the last time that I managed to take a holiday by myself and the aim was to see some more of the working between Newton Abbot and Plymouth, but most important of all the Axminster and Lyme Regis branch of the Southern. I started off early to spend an hour at Steventon and this picture is of No. 70024 **Vulcan** with the up *Capitals United Express* on the fast stretch shortly before Didcot.

19 May 1958

WINCHESTER
CITY
82014

76 And then on to Winchester, still a very dull day, with a standard 2-6-2T No. 82014 passing through the city station. I wonder if the large rack of bicycles is still in use on the down platform?

19 May 1958

77 Unfortunately I was not here when City of Truro was working this line and on the day in question No. 76064, a standard 2-6-0, came off the G.W.R. branch at Shawford Junction, onto the relief line, with a train from Didcot to Southampton Terminus. In the distance above the tender of the engine can be seen Winchester Cathedral and over on the right St. Catherine's Hill.

19 May 1958

30028

78 We pass the sight of Stockbridge station every year when we go on holiday and it is difficult to believe that I took this picture so many years ago judging by the scene today. Class M7 0-4-4T No. 30028 stops with a push-pull set on its way from Andover.

19 May 1958

79 After an overnight stay in Salisbury I went to the Tunnel Junction to see the early morning express leave for Waterloo headed by No. 34009 **Lyme Regis.** The line in the left foreground, complete with check rails for the sharp curve, comes in from Southampton.

20 May 1958

80

This will be a semi-fast to Basingstoke and beyond, taken from the A30 road bridge and looking north-east. 'King Arthur' class 4-6-0 No. 30452 **Sir Meliagrance** makes its way into the misty morning.

20 May 1958

81

Class N 2-6-0 No. 31813 comes under the road bridge with a train of weed killer wagons, a guard's van being inserted at each end.

20 May 1958

31813
SPL
15

73050
73050
DZ 6480

82 I should like to think that this is the Pines Express on its way from Bournemouth but with the time at around 11.30 I feel it is unlikely. Anyway, it makes a good picture as it passes Henstridge station hauled by standard class 5 Nos. 73050 and 73051. Fancy scrapping that nice Austin 7 pick-up!

20 May 1958

83 Motoring on to Sherborne, I took three pictures from the roadbridge carrying the A352 into the town. This must be a Salisbury to Exeter stopping train and hauled by 'Schools' class No. 30903 **Charterhouse.**

20 May 1958

84 With the 'Schools' class disappearing into the distance, the 10.30 from Exeter to Waterloo approaches Sherborne behind No. 34072 **257 Squadron.** The up line is about to be re-laid and you can see the new flat bottom rail lying along the edge of the sleepers.

20 May 1958

85 I expect experts on Southern working can identify this train which has just pulled away from Sherborne station on its way to Exeter. No. 34011 **Tavistock** has a load of 12 coaches and a 4-wheel parcels van.

20 May 1958

34011

86

A meeting outside the south portal of Crewkerne Tunnel, another example of the game of luck. The coal allocation in No. 34029 **Lundy** seems to be a bit on the spartan side and No. 34078 **222 Squadron** has steam to spare.

20 May 1958

87

And now we come to the object of the exercise and what a glorious sight it is—Adams Radial Tank No. 30582 in polished black livery and one coach. My favourite branch line from Axminster to Lyme Regis—alas no more. The train is about one mile out of Axminster.

20 May 1958

30582
30582

30582
10

88 I went to a different spot for the next return working in the afternoon but again not far away from Axminster and as you can see the day has turned out to be magnificent.

20 May 1958

89 Between the workings of the branch I went to the foot of Honiton Bank and caught this 'Merchant Navy' class No. 35029 **Ellerman Lines** on a midday train from Waterloo to Exeter.

20 May 1958

30582

90 The last trip down the branch in the evening was hopefully to have a request smoke effect but it did not produce much to liven up the picture. Note the cans of oil above the buffer beam, and the engine is just at the foot of the incline which will take it up and over the mainline to head down towards the south coast.

20 May 1958

91 No. 7017 **G. J. Churchward** leaves Totnes with the 07.30 Truro-Paddington and running without any assistance over the Devon banks to Newton Abbot. What a profusion of telegraph poles and wires on the up side of the line, and note the tarpaulin on the side of the goods shed keeping out the weather. A 2-6-2T waits in the station for the next freight to bank up Rattery incline, see picture No. 18 in *Reflections of the Great Western*.

22 May 1958

6004
BAY TOR
BOARDING KENNELS

92 Sweeping round the curves into Teignmouth, No. 6004 **King George III** is working the 09.40 Falmouth-Paddington. Rather a cramped sight for photography but the magnificent sky and clean engine help the picture. The chimney has now been demolished.

26 May 1958

93 *The Shamrock* was one of the three titled trains to serve the London-Liverpool route and this is the up train on a Saturday morning when it left Liverpool at 08.20 and arrived at Euston at 12.18. The motive power is No. 46210 **Lady Patricia** in green livery and it is seen near Stableford.

14 June 1958

94 And so we finish this Volume with another 'Princess Royal' class locomotive, but in maroon livery and soon after passing Whitmore. No. 46207 **Princess Arthur of Connaught** carries the headboard of *The Merseyside Express* departing Liverpool 10.10 and arriving London 13.45.

14 June 1958

Reflections of the BIG FOUR

R. J. BLENKINSOP

1 Whitmore station is visible in the background as No. 46256 **Sir William A. Stanier, F.R.S.** nears the summit of the climb out of Crewe and heads for Stafford.

14 June 1958

W478
46115

2 Looking as if it has recently been through Crewe works No. 46115 **Scots Guardsman** approaches Betley Road signal box and on towards London. The engine is now preserved at Dinting.

14 June 1958

3 In the background can be seen the bridge carrying the Newcastle under Lyme to Market Drayton railway, with the platforms of Madeley station in the foreground. Class 5 No. 44770 is on the down fast heading for Crewe.

14 June 1958

W270
46163

4 The fireman looks out ahead for the next set of signals and No. 46163 **Civil Service Rifleman** is working hard with a 13-coach load on the climb to Whitmore. The cows on the bridge are unperturbed as they go home for milking.

14 June 1958

5 Betley Road signal box with two down trains and on the up slow line a Class 5 heads south — rather a lucky shot. Note the different liveries of the coaches and the full sidings. On the left is No. 46205 **Princess Victoria** and Class 5 No. 45020 on the right.

14 June 1958

6 With a very clear exhaust No. 6020 **King Henry IV** climbs under the road bridge near the top of Hatton Bank. The train is the 17.10 from Paddington to Wolverhampton, the engine having worked up to London leaving the Black Country at 11.35.

8 July 1958

7 Photograph No. 88 in *Shadows of The Big Four* is a view from this same road bridge but the other side of the line. No. 45734 **Meteor** accelerates out of Coventry with the fast for Euston leaving the motor city of the Midlands around 08.00 in the morning.

23 July 1958

45734

8 Quite a crowd of locomotive spotters watch the summer working as the trains pass under the city walls of Chester. Class 5 No. 75013 of British Railways design approaches the city and another train leaves on its way along the north Wales coast.

9 August 1958

9 Stanier 2-6-0 No. 42965 leaves Chester with a holiday extra for Bangor. Note the motor cyclist with a crash helmet which was quite unusual in those days.

9 August 1958

W457

W49
45736

10

This is one of my favourites as it shows a relief to the *Irish Mail* passing the city walls on the slow line. The attraction is the lighting with the exposure calculated for the shadows of No. 45736 **Phoenix,** one of the two 'Jubilee' class engines rebuilt with double chimney and large boiler.

9 August 1958

11

And here we get nearer home, just over the fields from where I live. On a stormy morning a lucky shaft of sunshine catches No. 5019 **Treago Castle** as it climbs out of Leamington Spa with the 10.00 from Birmingham Snow Hill.

9 September 1958

12 I am sorry about the telegraph wires but that should not detract from the splendid sight of No. 5091 **Cleeve Abbey** with the down *South Wales Pullman.* It is shown here approaching Goring station on a very misty morning.

12 September 1958

13 The LNWR 0-8-0 on the left, works the exchange sidings for Keresley Colliery and on the mainline No. 49441 approaches Foleshill station with a coal train from Nuneaton heading for Coventry.

7 April 1959

14 The next day I was up early to see one or two trains near St Neots on the Eastern Region. The first was the 08.20 Kings Cross to Doncaster travelling very fast as it approached St Neots station behind No. 60010 **Dominion of Canada**.

8 April 1959

15 In the evening I was near Hatfield watching the succession of trains leaving London for the north. B1 No. 61082 makes for Cambridge on the mainline and a local stopping train approaches on the relief line. The coaches are in maroon livery and the locomotives black.

8 April 1959

61082

60862

16 Perhaps the cleanest engine was this V2 No. 60862 with a fitted freight and just in time before the sun disappeared behind the clouds on the horizon.

8 April 1959

17 Another day out on the Eastern Region included a visit to Stoke Bank to the south of Grantham. Here is No. 60034 **Lord Faringdon** climbing the gradient with the down *Flying Scotsman*.

23 May 1959

18 When you see the filth coming out of the chimney of a steam engine it is not surprising that they required a lot of regular cleaning. Many of the sheds did not have the staff and this V2 No. 60893 looks to be in a sorry state as it climbs Stoke.

23 May 1959

19 At the top of the gradient is Stoke tunnel and A1 'Pacific' No. 60125 **Scottish Union** emerges into the daylight with a northbound express.

23 May 1959

20 This is the scene on the climb up to Stoke from Grantham where a V2 has just failed on an up goods train. As the relief engine No. 60050 **Persimmon** backs gently down to couple up, a fast goes by for London behind No. 60055 **Woolwinder.**

23 May 1959

21 Signals are off in both directions as No. 60085 **Manna** comes past the iron ore exchange sidings outside Stoke tunnel and on its way to Peterborough.

23 May 1959

22

This is just outside Grantham near Great Ponton and two 'Pacifics' are seen passing each other in the late afternoon sun. In the foreground is No. 60150 **Willbrook** and an unidentified A3 'Pacific' with three coaches on its way south.

23 May 1959

23

No. 60017 **Silver Fox** has a down freight train as it passes a break in the trees allowing the sun to shine on the top part of the locomotive. Note the Silver Fox on the boiler cladding behind the nameplate.

23 May 1959

24 And now we come to the purpose of the journey, to see the *Stephenson Locomotive Society Jubilee Special* make the ascent to Stoke. Here it is at 75 mph climbing the 1 in 200 gradient behind No. 60007 **Sir Nigel Gresley** which was to reach 112 mph near Little Bytham.

23 May 1959

25 It is quite appropriate to have a 'Coronation' class Pacific on the opposite page to an LNER A4. At this time *The Caledonian* was being worked by No. 46245 **City of London** and here it is approaching Brinklow Station shortly after passing through Rugby at reduced speed.

3 June 1959

THE CALEDONIAN
46245

26 The next four pictures were taken within a short space of time at Victoria Station in London. 'West Country' class No. 34103 **Calstock** is shown just before departure.

29 July 1959

27 In this picture another train has just left and the banker can be seen hard at work. Note the fashion in the ladies clothes and the brackets for taking the *Golden Arrow* boards on the side of the engine.

29 July 1959

28 **Calstock** has now left the station and is being assisted up the gradient by 'Schools' class No. 30921 **Shrewsbury.**

29 July 1959

29 Awaiting the departure of *The Golden Arrow* is No. 35028 **Clan Line** looking very smart in unrebuilt condition and of course polished to perfection. It is nice to know that it is in preservation at Hereford and comes out from time to time in the summer months.

29 July 1959

35028

CHELTENHAM SPA
EXPRESS
502
215

30 Before taking my train back to Leamington Spa I took a few pictures at Paddington and this one shows No. 5094 **Tretower Castle** leaving with the titled train for Cheltenham and Gloucester.

29 July 1959

31 Photograph No. 43 in *Echoes of the Big Four* shows the other side of this flyover at Weaver Junction. In the afternoon a northbound express hauled by No. 46157 **The Royal Artilleryman** has just passed under the up mainline from Liverpool.

29 August 1959

M985
1000
1000
M R

32

When the Midland Compound was preserved it ran a number of Specials for Locomotive Societies and this one ran from Birmingham New Street to Doncaster to see round the Locomotive works. Organised by the Stephenson Locomotive Society it was stopped by signals just where I was waiting to take a picture at Mexborough. Held for some ten minutes by signals there was time to allow the driver to check round the engine and see that all was in order.

30 August 1959

33

A self-weighing tender is just visible on the left and outside after receiving its final coat of paint is No. 60090 **Grand Parade**. My first and only visit to the Doncaster Works.

30 August 1959

34 Black engines and black sheds do not make easy pictures but 2-8-0 No. 63948 is just ex-works and the chances of seeing it again so clean are very remote. Taken at Doncaster Sheds.

30 August 1959

35 In the afternoon the train continued to York for a visit to the Railway Museum, not the one we know today. I suppose this could happen today with the same locomotive by visiting the National Railway Museum and by a different route.

30 August 1959

M985
1000
1000
M R
DN60

7801

36 I must have been on holiday early in September as this one is taken at Machynlleth just two days later. No. 7801 **Anthony Manor** has just arrived in the station with the freight from Aberystwyth and will shortly be leaving for Shrewsbury with a banker for the climb to Talerddig.

1 September 1959

37 The driver is looking ahead and the fireman is about to swing a shovelful of coal into the firebox of No. 46116 **Irish Guardsman** as this London to Holyhead express has just crossed the River Dee outside Chester.

2 September 1959

W337

38 The speed restriction sign showing a maximum speed of 50 mph for the points at Weaver Junction is just visible at the left of the picture. No. 45580 **Burma** has crossed the River Weaver with a down express.

5 September 1959

39 In the opposite direction and taken from the same spot No. 46243 **City of Lancaster** hurries by with the up *Royal Scot,* the engine being in maroon livery.

5 September 1959

40 Another special, this time for the Talyllyn Railway Preservation Society from London to Towyn for the Annual General Meeting. 'Dukedogs' Nos. 9004 and 9014 speed down the valley from Cemmaes Road to Machynlleth.

26 September 1959

41 Cup Final day at Wembley and as Wolverhampton Wanderers were in the match a number of specials were run. This one is taken near Claydon Crossing to the north of Banbury and the tender of No. 6006 **King George I** has been suitably decorated.

6 May 1960

GOOD LUCK
WOLVES

7F
53807

42 My only visit to the Somerset and Dorset was to see a special hauled by S & D 2-8-0 No. 53807 leave the city of Bath. Here is the engine at Bath Green Park Shed after being prepared for the run. Note the tablet exchange apparatus on the tender.

14 May 1960

43 No. 40634 was originally built for the Somerset and Dorset whilst No. 40700 was the last of the class built for the LMSR. Both engines of course were made after grouping and it looks as if 40700 is out of use due to the chimney being covered over.

14 May 1960

44 In a corner of the Shed at Bath was this delightful Johnson design (Midland Railway) 0-4-4 tank engine awaiting its fate.

14 May 1960

45 Now here is a sight worth waiting for as No. 53807 comes up the gradient out of Bath in the glorious sunshine. Note the fire irons on the top of the tender and the clean design of this engine.

14 May 1960

53807
W 9349

395
44557
40

46 Following the special was the *Pines Express* and after that this local train stopping at all stations. The 0-6-0 No. 44557 looks a bit grimy but the coaches are just out of the works in Southern green. The location is of course to the south of Midford Station.

14 May 1960

47 On the way home I stopped in the outskirts of Bath to see this express arrive behind No. 6932 **Burwarton Hall.** It is approaching the city from the London direction and is probably a holiday extra from the south.

14 May 1960

48 The final call for the day was at Stratton St Margaret outside Swindon and the daily running in turn is coming in from the east behind 2-8-0 No. 2892.

14 May 1960

49 An historic view of the rebuilding of Coventry station. With a local train in the down platform the tank engine and coach were used to take parcels from one side to the other. The picture is looking south and I like the two men chasing a copy of the daily paper across the track!

3 June 1960

50 We had our summer holiday in North Wales and although not built to mainline gauge the Festiniog has been a favourite for many years and I cannot help including this one. The bridge in the background had just been rebuilt and necessitated a temporary level crossing shown to the right of the picture. **Taliesin** approaches Boston Lodge with a return afternoon train.

7 June 1960

51 The 07.33 off Pwllheli has just left the station and I assume it is on its way to Bangor. With a handsome looking 2-6-4 tank engine No. 42074 it makes a fine sight in the early morning.

10 June 1960

48686

52 The place Leamington Spa GWR shed. The engine a Stanier 2-8-0 just received back from Crewe works after overhaul. Alas it must have run a hot big end as the connecting rod is lying on the ground protected by a couple of lamps.

19 June 1960

53 The next four pictures were taken on the visit o Crewe to see how the electrification was progressing. The *Pines Expiess* approaches behind a very dirty 'Britannia' class Pacific No. 70042 **Lord Roberts** from the Manchester end.

25 June 1960

HOME GUARD

54 At the south end of the station one of the last 'Patriot' class engines in service has arrived with a passenger train. The *Merseyside Express* runs through behind No. D.255.

25 June 1960

55 I saw this engine leave Crewe for Chester in the morning so thought it would come back in the afternoon. A request to the fireman for smoke during a chat at Chester did not provide much black on the troughs outside the city.

25 June 1960

56 The fireman leans out of the cab of No. 46241 **City of Edinburgh** as water is picked up by the scoop under the tender.

25 June 1960

57 20.00 hrs on a summer evening at Liverpool Street Station with one aim in view – to see the two immaculate station pilots. 0-6-0 No. 68619 in Great Eastern livery makes interesting contrast to No. 70040 **Clive of India** with a boat express.

28 June 1960

68619
70040
THE HOOK CONTINENTAL

69614
68619

58 And here they are together with the driver of No. 69614 oiling up the motion. I gathered from conversation with the crews that the reason the engines were kept so clean was that extra payment was given.

28 June 1960

59 And so to the final picture with quite a line up at 20.35 in the evening. Perhaps I can leave you to work out the classes of all the engines.

28 June 1960

V05

60 I came back to London a few days later behind No. 6006 **King George I** on the 11.35 Wolverhampton to Paddington and here is the train after arrival.

30 June 1960

61 Down at Old Oak Common an Austerity 2-8-0 has the fire lit up and clouds of acrid smoke are shown coming out of the chimney. Was it the dirt and long hours of work which killed the steam engine? No, I believe it had served the purpose for which it was designed for well over a hundred years and died a natural death from our advance in technology. You cannot go backwards or stay the same but only progress to better things.

30 June 1960

EARL OF MOUNT EDGCUMBE
92249

62 Scenes such as this make romantic viewing and memories linger on but I am sure no one would wish to return to the darkest corners of Old Oak Common Shed.

30 June 1960

63 No. 3440 **City of Truro** climbs out of Leamington, wrong line working, with a Stephenson Locomotive Society Special from Birmingham to Swindon. Quite a crowd on the footplate!

4 September 1960

R.C.T.S.
THE
EAST MIDLANDER
1000
1X80V
M R
1000

64

A week later the Midland Compound was in action on a special from Nottingham to Swindon via Reading, Basingstoke, Andover and up the MSWJR. The Compound came off at Oxford, I believe, and ran light to Swindon to await the return trip. Here is the train on Aynho water troughs to the south of Banbury on a cloudless day with no exhaust.

11 September 1960

65

Now down-graded and single track from Princes Risborough to Aynho Junction, the shortest route by the Western Region to Birmingham was used to capacity eighteen years ago. In this shot No. 6017 **King Edward IV** with steam to spare approaches Beaconsfield with the 14.10 Paddington to Birkenhead.

17 September 1960

66 Another running in turn which was a regular feature at Crewe works was an early morning train stopping at all stations to Shrewsbury. In this view No. 46245 **City of London** is nearing the outskirts of Shrewsbury on a misty morning.

24 September 1960

67 In chalk at the bottom of the smokebox of No. 4918 **Dartington Hall** is written 'NOT TO BE BLOWN' and I wonder what this can mean. The scene is of course Shrewsbury looking south with Class 5 No. 45395 awaiting departure for Crewe.

24 September 1960

REWSBURY
SHREWSBURY
4918

68 Another engine just out of Crewe works is this Class 5 No. 45448 seen at Shrewsbury shed in the evening awaiting its next call of duty.
24 September 1960

69 This route from Shrewsbury to Hereford and South Wales rarely sees trains as long as this one today. Hauled by No. 5044 **Earl of Dunraven,** it is near Stapleton and will soon be starting the climb up to Church Stretton.
24 September 1960

V58

70 Another favourite spot half way down Hatton Bank as the up trains have usually shut off steam and a nice trail of black smoke flows along the top of the coaches. No. 6017 **King Edward IV** has the 11.00 from Birmingham Snow Hill to Paddington.

4 February 1961

71 Knowing that No. 46203 **Princess Margaret Rose** was in Crewe works for overhaul I arranged to be allowed to take some photographs of the engine after it emerged from the paint shop. Here it is ex-works for the last time in green livery and it is good to know that it survives today.

7 February 1961

72 Thinking back over the years it is not always easy to locate the exact spot where some of the pictures were taken. At the back of Edge Hill Shed in Liverpool 0-8-0 No. 49002 passes by with a freight train heading for the city.

18 February 1961

73 I then went down to Liverpool Lime Street Station to see a well groomed 'Scot' leaving on a train to Chester. No. 46163 **Civil Service Rifleman** simmers away underneath the overall roof of the station.

18 February 1961

46163

74 In the afternoon I went to Chester and the state of these two engines of similar class is interesting, watched by engine spotters of the day. 2-6-4 tank engine No. 42183 on the left and No. 42583 on the right.
18 February 1961

75 With less than a year of its life left before withdrawal No. 6006 **King George I** climbs out of Harbury Tunnel with the 11.00 Birmingham Snow Hill to Paddington. This engine was the first of the 'Kings' to be withdrawn in February 1962.
29 March 1961

V03

76 Cup Final day again but this time all the specials came down the Great Central mainline. No. 46160 **Queen Victoria's Rifleman** has just come through Catesby Tunnel after the climb from Rugby. Go and have a look at the scene today.

6 April 1961

77. I am standing by the bridge which carries the Stratford-upon-Avon and Midland Junction Railway over the Great Central. 2-6-4 tank engine No. 67740 heads south and will most likely take the branch to Banbury.

6 April 1961

2211
GEORGES
PRIVATE
KHP 268

78 The approach to Leamington Spa from the south as seen by the railway traveller used to be pretty grim. This picture shows some of the demolition taking place at that time for a new Industrial Estate to be built. No. 2211 runs past the signals set for entry to the Great Western Shed. I wonder if the coach was preserved!

12 April 1961

79 The last year of working *The Inter-City* by steam finds No. 6029 **King Edward VIII** leaving Leamington Spa on a dull evening. The scene has altered dramatically today.

26 May 1961

80

My Southern content really started when we took to the Isle of Wight for holidays each year and like many others I fell for the Island railways. In 1961 Ryde Pier Head to Cowes and also Ventnor were still in full swing with the engines in a reasonable state of cleanliness. The 18.30 from Ryde to Cowes has just left Small-brook Junction with the next stop Ashey Station.

20 June 1961

81

A broadside view of an O2 at speed approaching Ryde St John's Road Station with a train from Ventnor and hauled by No. 22 **Brading.**

21 June 1961

22
BRADING

82 A few minutes later No. 25 **Godshill** appears with a coal train from Medina Wharf up the river near Cowes. This will be for the locomotive depot at Ryde.

21 June 1961

83 The coal train has now been split up and it is seen on the left of the picture. Passing by on the 18.25 from Ryde to Ventnor is No. 29 **Alverstone.**

21 June 1961

MCW 54

84 One of the routes to the Isle of Wight is Lymington to Yarmouth and here the cars are queueing for the ferry. A nice collection of vehicles but the push-pull from Brockenhurst should be the attraction, hauled by 0-4-4 tank engine No. 30133.

24 June 1961

85 I spent a Saturday in the Brockenhurst area when all the summer service trains were running. Here No. 30781 **Sir Aglovale** is approaching Brockenhurst Station with the 09.30 Bournemouth to Wolverhampton.

24 June 1961

60
34037
255
386

86 It turned out to be a cloudless day and very hot with trains going past every few minutes. No. 34037 **Clovelly** has just passed Brockenhurst with the 10.15 Bournemouth to Waterloo.

24 June 1961

87 The 'Schools' class 4-4-0 express engines were still running and one or two were on the trains for the Isle of Wight. No. 30927 **Clifton** is slowing down for the Brockenhurst stop with the 9.55 Woking to Weymouth.

24 June 1961

30853

30855

88 No. 30853 **Sir Richard Grenville** heads for Southampton with the 13.05 Bournemouth to Waterloo.

24 June 1961

89 And slowing down for the Brockenhurst stop is No. 30855 **Robert Blake** with the 12.20 Waterloo to Bournemouth.

24 June 1961

90 With the sun high in the sky No. 34016 **Bodmin** approaches the signals which have just been cleared. The train is the 14.05 from Eastleigh to Bournemouth and as you probably know the engine is in preservation on the Mid-Hants line.

24 June 1961

91 Later in the afternoon I took a train to Beaulieu Road Station and walked into the New Forest to see the trains. No. 35017 **Belgian Marine** heads for London with the up *Bournemouth Belle.*

24 June 1961

92 No. 34103 **Calstock** passes in the other direction with the 10.00 Bradford Exchange to Poole.

24 June 1961

93 Back to Brockenhurst in the evening, No. 34077 **603 Squadron** approaches the station at speed with a returning excursion.

24 June 1961

MERCHANT
NEW ZEALAND
LINE
NAVY CLASS

94 Whilst waiting for my train to Lymington No. 35021 **New Zealand Line** stopped at the station with the 17.30 from Waterloo to Bournemouth. No. 30133 arrives from Lymington.

24 June 1961

95 Back on the Isle of Wight No. 16 **Ventnor** approaches Smallbrook Junction with the 08.25 Ryde to Ventnor train. Note the fixed distant signals on the road bridge.

28 June 1961

96 And for the last picture in this volume No. 21 **Sandown** is coming down the bank from Smallbrook Junction to Ryde St John's Road station with an evening train from Ventnor. Just for once I failed to record the date!

SILHOUETTES OF THE BIG FOUR

R.J. BLENKINSOP

1 A dull July morning sees No 6016 **King Edward V** passing Old Oak Common and taking the Birmingham route, through Bicester, with the *Inter-City*. On the right is a pannier tank with a local freight waiting for the road.

26th July 1961

48142

△

2 Stanier 8F No 48142 crosses the Western Region main line near North Acton with a transfer freight for the Midland Region.

26th July 1961

3 Taken from the same location an unusual picture of V2 No 60879 climbing to join the Western Region with a train that it has most probably taken over at Kensington Olympia. This double track has now been lifted and the London Transport lines in the foreground are part of the Central Line.

26th July 1961

WEST RUISLIP

△

4 Here we are looking south with Wormwood Scrubs Prison behind 'Britannia' Pacific No 70016 **Ariel** awaiting signals with a parcels train for Paddington. In the foreground the tube train makes for West Ruislip.

26th July 1961

5 The location is Fenny Compton and shows in the foreground the track of the former Stratford-upon-Avon and Midland Junction Railway looking towards Woodford Halse. No 6029 **King Edward VIII** has just slowed for a P. W. restriction with the down *Cambrian Coast Express.*

29th July 1961

6 I had a few minutes to spare in Huddersfield one day and caught these two coming into the station with mainly E.R. stock. 2-6-2 tank No 42414 pilots 8F No 48265.

23rd August 1961

7 On the way home, and after passing Buxton, I came to Parsley Hay on the former Ashbourne line. Ex L.N.W.R. G2a 0-8-0 No 49391 leaves the station for Buxton. This delightful spot today is preserved for country walking and cycling.

▽

23rd August 1961

46158
BRITISH LWAYS
BD 48419B

△

8 Nearer home at Leamington Spa there was an evening freight working which came through from Banbury and was always headed by a Midland Region engine. At that time a small industrial estate was being built and it was possible to climb on to the roof tops and, in this case, see No 46118 **Royal Welch Fusilier** gently easing forward as the driver waited for the signals to clear.

31st August 1961

9 A dirty locomotive with maroon coaches on a misty morning might not seem to make a good picture. However railways often looked like this and I expect the awaiting passengers at Princes Risborough would not be really interested providing B1 No 61234 reached Nottingham on time.

2nd September 1961

10 Back to the industrial estate at Leamington Spa to see the *Inter-City* leave on its way to London double headed by No 7026 **Tenby Castle** and No 5032 **Usk Castle**. The bricklayer at least appreciates the sight and sound!

22nd September 1961

11 It must be Kings Cross and Deltic No D 9011 **The Royal Northumberland Fusiliers** just one month old poses beside Gresley A4 Pacific whose name and number I am sure you can read!

▽

25th September 1961

ENQUIRIES
TICKET OFFICE
DOMINION OF NEW ZEALAND
D9011

12 Just to give prominence to the steam world in face of the diesel invasion, here is another picture taken from the west side of the station but the attraction is of course the human interest and the advertising.

25th September 1961

13 The Pullman coach carries the board 'Birmingham Pullman' and No 6020 **King Henry IV** reaches the end of its journey with the morning train from the Midlands to Paddington.

25th September 1961

82006

14 On the Merionethshire Coast at Towyn a B.R. Standard Class 2-6-2 Tank Engine No 82006 leaves for Machynlleth with a local train from Pwllheli. Note the simple layout of the yard and the track on the left with concrete supports for each chair.

30th September 1961

15 With a Crewe North Shed plate in evidence and looking very clean I am sure 2-6-0 No 78030 has recently been through the works. The locomotive is acting as station pilot at Crewe and is reversing through the station.

4th November 1961

46154

16 At the north end of the station the local enthusiasts are watching the arrival of No 46154 **The Hussar** with a train from Chester.

4th November 1961

17 It is good to be reminded of the signalling at Leamington Spa as it is all colour light today. No 6000 **King George V** awaits the green flag with the down *Cambrian Coast Express.*

11th November 1961

730
7014

18 The small frame in front of the fogman's hut is used for placing the detonators on the down line under the centre of the leading coach. No 7014 **Caerhays Castle** breasts the top of Hatton Bank with a football special.

17th February 1962

19 Steam and Diesel at Kings Cross with the station staff cleaning carriage windows. 'Britannia' Pacific No 70038 **Robin Hood** has just arrived from Cleethorpes and stands in a ray of sunshine with the arch of the station framing the picture.

14th March 1962

67793
IA 30

20 I failed to record the name of the Deltic about to leave Kings Cross but the enthusiast seems to be looking it up in his book. Class L1 2-6-4T No 67793 makes a nice contrast alongside.

14th March 1962

21 No 5991 **Gresham Hall** passes through Swindon station with a west-bound freight. The photograph makes an interesting comparison with the very changed scene of today.

31st March 1962

68723
XP
68723

△

22 A quick visit to Newcastle by train produced this picture taken out of the carriage window as the southbound journey commenced. Class J72 0-6-0 tank engine No 68723 acts as station pilot and is painted in North Eastern livery.

20th April 1962

23 *The Aberdeen Flyer* was a special train from Kings Cross to Aberdeen and back down the west coast to Euston. Running over two hours late, due to electrification of the line, No 46200 **The Princess Royal** leaves Rugby with the familiar wireless masts in the background.

3rd June 1962

24 Now safely in preservation at Dinting No 45596 **Bahamas** heads north on the West Coast main line near Cranberry, north of Stafford. The wires are up but electric haulage had not yet commenced.

9th June 1962

25 In the opposite direction and travelling very fast, 'Britannia' Pacific No 70046 **Anzac** has the driver looking very carefully from the cab window.

▽

9th June 1962

ANZAC
70046

26 In the afternoon I moved north to Whitmore and saw No 71000 **Duke of Gloucester** coming round the curve from Crewe. This is the engine now being restored at Loughborough. Note the Caprotti valve gear.

9th June 1962

27 On the water troughs now, and unnamed 'Britannia' Pacific No 70047 overfills its tender quite surprisingly as it is near the commencement of the pick up.

▽

9th June 1962

70047

28 No 46240 **City of Coventry** passes the sight of the former station at Whitmore and the steam coming from the rear of the tender shows that the coal pusher is in operation. A fine looking locomotive travelling at speed.

9th June 1962

29 Back to Swindon again with a view at the west end of the station as No 6931 **Aldborough Hall** is inspected by crowds of engine spotters on all platforms. The works is of course in the background.

▽

13th June 1962

6931

30 While the Euston–Crewe electrification was being carried out Sunday trains came through Leamington Avenue station on various occasions. *The Ulster Express* with No 46208 **Princess Helena Victoria** passes on its way north and the W.R. station is in the background.

1st July 1962

31 I much enjoyed the occasional visit to Kings Cross as there was a total mix of Steam and Diesel. 'Britannia' Pacific No 70038 **Robin Hood** eases its train into the station.

▽ **18th July 1962**

S
70038

CAMP
Shapes
60134
FOXHUNTER
1A 09

32 The mixture becomes more apparent down at the buffer stops with the diesel even having brackets above the buffers for taking oil lamps!

18th July 1962

33 A departure with No 60119 **Patrick Stirling** just about to slip violently but too late for my camera.

18th July 1962

60029
WOODCOCK
60029

△

34 The driver of No 60029 **Woodcock** moves slowly down to the coaches as the fireman prepares to couple up having first removed the lamp which he has placed on the ground.

18th July 1962

35 The heap of coal on the tender looks as if it is outside the loading gauge. The diesel-hauled express has a set of Pullman cars with one of the conductors leaning out of the first coach.

18th July 1962

36 An unusual combination at the arrival side of Paddington Station with No D854 **Tiger** leading No 4932 **Hatherton Hall**. Up in Brunel's roof tarpaulins hang to protect the passengers from the painters!

18th July 1962

37 The next set of pictures were taken around Southampton during the summer timetable. No 34059 **Sir Archibald Sinclair** awaits departure with S15 Class No 30508 in the background.

▽ **25th August 1962**

SIR ARCHIBALD SINCLAIR
7P 5FA
34059

38 The train number on the smokebox has covered the locomotive number so I cannot tell you which one it is. Note the design of the station lamp brackets.

25th August 1962

39 This is the classic view at Southampton showing the signal gantry and docks in the background. North Eastern coaches abound as they approach the station behind No 34041 **Wilton.**

▽

25th August 1962

40 The sunlight near the coast is always brilliant and No 34050 **Royal Observer Corps** glistens as it departs for Bournemouth.

25th August 1962

41 Nos 41329 and 82014 approach Totton with an oil tanker train from the Fawley Refinery. Note the two wagons inserted to reduce any fire risk from the locomotives, and I wonder if the drivers are having an interesting conversation.

▽ **25th August 1962**

82014
41329

42 In the afternoon the return working for Waterloo started to appear and this one has come up from Lymington with passengers from the Isle of Wight. 'Schools' Class No 30935 **Sevenoaks** emerges from under the A35.
25th August 1962

43 Passing Redbridge signal box is another up express from Bournemouth with No 34032 **Camelford** in charge. Note the small shed with engine off duty for the weekend.

34032
385
30
40

44 Taken from the same footbridge but showing a train coming in from the Salisbury line and most likely going to Brighton. No 34055 **Fighter Pilot** emits a healthy heat haze in front of the cottages.

25th August 1962

45 Just to the west of Southampton station the line swings round to the north west and No 34102 **Lapford** passes a well cared for Austin Seven standing in the brilliant sunshine.

25th August 1962

▽

34102
039

46 This was the one I had been waiting for all day and for once the sun shone. The picture of this magnificent sight tells all and I only need to mention that the locomotive is No 35030 **Elder Dempster Lines.**

25th August 1962

47 Most trains took water at Southampton and there was time for me to run back to the station and see the *Bournemouth Belle* having the tender refilled. In the foreground is No 34006 **Bude** and a delightful collection of spotters.

▽

25th August 1962

SOUTHAMPTON
CENTRAL
34006
34006
301
36
35030
33

48 Perhaps three pictures of the same train are rather too many but I could not resist this one of the departure with a close up of the first car with passengers sitting at their tables. Note the spectators above the tunnel mouth.

25th August 1962

49 Most steam enthusiasts enjoy the sight of a paddle steamer and this one was built for the Southern Railway in 1937. Here it is arriving off Ryde Pier Head with some holidaymakers from Southsea.

August 1962

SANDOWN

△

50 This could be called a 'conversation piece' at Hayling Island Station. Note the size of 'Terrier' Class No 32661 compared with the coach, and the island platform with gas lighting.

27th August 1962

51 Langstone Bridge, passed shortly after leaving Havant, was an all timber affair and here the 'Terrier' is on its way south. Notice the road bridge in the background.

27th August 1962

52 The two Class N Moguls Nos 31871 and 31868 give a Southern Region atmosphere to this scene at Reading General. No 5058 **Earl of Clancarty** passes through with a returning train of milk wagons.
14th September 1962

53 This was the year when the Western diesels were introduced in large numbers and here we see No D1005 **Western Venturer** climbing past Gresford Colliery with a Birkenhead to Paddington train, leaving Chester at 09.15.
▽
29th September 1962

1V04
D 1005

45268

△

54 A clean Class 5 No 45268 in the down platform at Leamington Spa. The present line to Coventry swings to the right just beyond the signal box.

20th October 1962

55 A few weeks later another Class 5 No 44712 is in trouble at the other end of the station with a steam crane preparing to lift the front end of the locomotive and replace it on the track.

18th November 1962

56 Now we move on to 1963 when No 4472 **Flying Scotsman** had been purchased for preservation and was making its first run in L.N.E.R. livery. A terrible day but it makes an impressive sight climbing towards Ruabon with the Festiniog Railway Preservation Society Special.

20th April 1963

57 In the afternoon the locomotive worked light engine to Shrewsbury and here it is crossing the River Dee.

▽

20th April 1963

58 On arrival at Shrewsbury **Flying Scotsman** was serviced and came onto the turntable. The return working left in the early hours of Sunday morning for Paddington.

20 April 1963

59 Southampton Football Club were playing in Birmingham and no less than thirteen special trains conveyed the supporters to Birmingham. All but one had S.R. engines and the majority climbed Hatton Bank in the morning. Here is No 34045 **Ottery St. Mary** at the location where many enthusiasts watch the special steam trains today.

▽

27th April 1963

34045
34045

34046
7929

△

60 In the afternoon there seemed more S.R. engines on Tyseley shed than G.W. and here No 34046 **Braunton** stands beside No 34039 **Boscastle** with a very subdued No 7929 **Wyke Hall** to the right of the picture.

27th April 1963

61 The following day No 6018 **King Henry VI** made the last run of the class having been withdrawn the previous autumn. Having attained a maximum speed of 92 m.p.h. near Denham the engine takes a rest on Swindon Shed amid Hymek hydraulic diesels, before returning to the Midlands.

28th April 1963

S·L·S
SPECIAL
Z77
6018

62 The route back to Birmingham was through Oxford and I wonder if this was the first occasion that a 'King' had worked through that city.

28th April 1963

63 Cup Final day at Wembley brought a number of Special trains down the Great Central route. Class 5 No 45334 is about to enter Catesby Tunnel near Charwelton.

25th May 1963

1X55
45626

64 At the other end of the Tunnel No 45626 **Seychelles** hurries through the cutting which is derelict and empty today.

25th May 1963

65 As this was the last main line built out of London the date on the tunnel mouth 1897 seems very modern. Class 9F No 92229 heads north for Rugby.

25th May 1963

△

66 Just to the south of Woodford Halse is a triangle connecting up with the S.M.J.R. and between the football specials an 8F No 48385 passes through with No 48002 in the background.

25th May 1963

67 The last special was headed by No 45598 **Basutoland** and is passing under the line from Stratford-upon-Avon to Blisworth.

25th May 1963

68 Further north at Staverton Road signal box I waited to see this up freight train hauled by 9F No 92072.

25th May 1963

69 An Autumn special train found No 45552 **Silver Jubilee** climbing the Lickey Incline with a 9F banker at the rear. The smoke effect was quite remarkable and the train went up to Birmingham and straight back again.

▽

12th October 1963

45552

45676

△

70 A service train runs into Barnt Green station behind No 45676 **Codrington** under the fine set of signals one of which has a wooden post.

12th October 1963

71 Another set of fine signals of L.N.W.R. origin are a well known feature of Chester Station. Class 5 No 45352 passes Chester No 4 Signal Box with a freight train from the Holyhead main line.

2nd November 1963

73040

△

72 The trains from Shrewsbury to Chester were often hauled by B.R. Standard locomotives and in this case Class 5 No 73040 comes into the arrival Bay.

2nd November 1963

73 The Stephenson Locomotive Society ran a railtour from Birmingham to Shrewsbury, Newport, Severn Tunnel Junction, Swindon and back to Birmingham. Here the train is seen from the carriage window slowly negotiating the west curve at Shrewsbury.

26th April 1964

S·L·S
SPECIAL
407

△

74 The locomotive for the tour was No 4079 **Pendennis Castle** and here the train is shortly to depart from Hereford having stopped for the 650 passengers to inspect Barton Shed.

26th April 1964

75 Southampton again with the fireman off No 34009 **Lyme Regis** either shouting to the driver to turn off the water or perhaps he is gazing at some attractive passenger on one of the other platforms!

6th June 1964

34004

△

76 Another 'conversation piece' this time at Southampton where a change of crew is taking place and refilling the tender provides time for both sets of men to exchange views on the mechanical state of No 34004 **Yeovil**.

6th June 1964

77 The Royal Train is in the exchange sidings at Leamington Spa conveying the Queen Mother from Kenilworth to Stratford-upon-Avon. This required a change of direction and another Class 5 is waiting to take over from No 45322.

11th July 1964

78 This departure from Aberystwyth on the Vale of Rheidol Railway shows No 9 **Prince of Wales** running along the old route past the engine sheds to be seen just above the fourth coach.

July 1964

79 My first and only visit to Eastleigh Works took place to coincide with a Warwickshire Railway Society special train from Birmingham behind No 4472 **Flying Scotsman**.

Two 'West Country' Pacifics stand in the sun outside the shed.

16th August 1964

80 Resplendent in a new coat of paint after overhaul in the works 'Terrier' tank No 32650 makes a stark contrast with the other larger engines. Fortunately it is now preserved on the Kent and East Sussex Railway.
16th August 1964

81 Within the works 'Merchant Navy' No 35012 **United States Lines** is receiving attention to the valve gear. Note the connecting rods lying on the floor.

35012
35012

30102
WELD

△

82 Another locomotive now in preservation at Bressingham Hall in Norfolk. Class B4 0-4-0T No 30102 is sitting quietly on two enormous timbers surrounded by the usual workshop bits and pieces.

16th August 1964

83 I make no excuse for including this one as it is one of my favourite shots of No 4472 **Flying Scotsman** as it leaves Leamington Spa with a special train for the Farnborough Air Display.

12th September 1964

84 A few days later was the annual Stephenson Locomotive Society trip to Swindon. In this case it was hauled by the last 'County' Class locomotive No 1011 **County of Chester** seen here at the east end of Swindon station.

20th September 1964

85 Down at the locomotive shed were two engines now safely in preservation. 9F No 92203 operates on the East Somerset Railway and No 7808 **Cookham Manor** is part of the collection at Didcot. On the left may be seen the buffers of M7 0-4-4T No 30667.

▽ **20th September 1964**

92203
7808
D7087

GREAT WESTERN
4555

△

86 Already destined for preservation No 4555 stands outside the Shed at Croes Newydd ready to take a special to Towyn.

26th September 1964

87 The shed lies on a triangle and here is an empty coal train taking the route to Brymbo coal mine. The locomotive 8F No 48090 passes by two A.T.C. test ramps in the yard.

26th September 1964

88 The special train for the Talyllyn Preservation Society has now left Ruabon and takes the line for Llangollen, Corwen, Dolgellau and Barmouth. The 'Manor' Class 4-6-0 No 7827 **Lydham Manor** is now in preservation on the Torbay Steam Railway.

26th September 1964

89 On my way to Neston I called in at Mold Junction Shed and found this trio resting peacefully in the late Autumn sunshine.

▽

26th September 1964

44913
42730
45395

90 'A last train' to run on the Stratford-upon-Avon and Midland Junction Railway was organised by the Stephenson Locomotive Society from Birmingham to Woodford Halse and return. In the outward direction 4F No 44188 and restored pannier tank No 6435 have just passed under the aqueduct at Bearley.

24th April 1965

91 The pannier tank only went as far as Stratford-upon-Avon and in this picture the 4F No 44188 is returning in the afternoon passing Byfield station.

S·L·S
SPECIAL
44188
44188

S·L·S SPECIAL
35017
SPL
7C
17

△

92 Another S.L.S. special *The Bulleid Pacific Rail Tour* ran from Birmingham to Exeter outward via Basingstoke and return through Westbury. No 35017 **Belgian Marine** stands at Exeter Central taking on water. It worked the leg from Salisbury to Westbury.

23rd May 1965

93 The last scheduled steam hauled service from Paddington ran to Banbury. No 7029 **Clun Castle** did the honours and here it is backing down to Banbury Shed with a 'Western' Class diesel hydraulic speeding past without stopping at the station.

11th June 1965

94 To end the volume, I have included this picture of a 9F No 92013 passing Charwelton village with a freight for the North Midlands. The water troughs are visible in the background—a desolate scene today.
Date unknown